Contact Info:

Email: rgolson532@gmail.com

About the Author

Ryan Olson is an 11 year old boy who has been very interested in astronomy since he was little.

Today, he is in the 6th grade and has lots of fun with his smart and funny friends. He also enjoys fishing, programming, reading books, doing math, and doing DIY projects.

DISCLAIMER

This information is provided at the reader's discretion. The author/ publisher are not responsible for any injuries caused by this book.

NOTE: Never look directly at the Sun! Always have adult

supervision and use special Sun filters for your telescope or binoculars if you are going to view the Sun.

For My Father

Who loves astronomy...

Table of Contents

1
What's that in the sky?

Have you ever wondered what happens above your home in space? Well, you're about to discover some of the coolest things in the entire universe!

In this book, you're going to discover fascinating things about our solar system and universe. You'll learn what happens in outer space, what tools scientists use to explore outer space, and the exciting world that most people don't realize is right outside their door above their heads! This book will teach you some very basic information about outer space and how easy it is to explore our solar system and universe. You're going to have a blast learning about what is in outer space!

2
What is our Solar System?

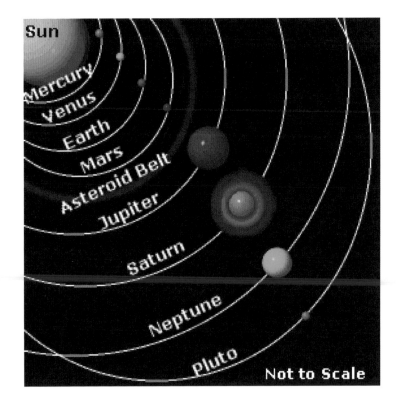

Our solar system is right outside of our Earth's atmosphere. It is kind of like a giant house that has the Sun, Mercury, Venus, Earth (and our Moon), Mars, Jupiter, Saturn, Uranus, Neptune, Pluto, and a bunch of other itty-bitty planets that we don't really call planets in it. The Sun is the center of our solar system. All the planets (including the Earth) and their moons revolve around our Sun. Cool, huh? It may seem that the Sun, is small at a glance, but it is really, really, REALLY big! It is so big you could fit over 1 million Earths inside it. That's pretty big!

WARNING!! Don't ever look at the Sun without a pair of Sun goggles specially made for looking at the Sun. You can seriously hurt your eyes and may become blind. So be careful! Be sure to ask your parents first before you look at the Sun through safety goggles! (See Disclaimer)

Some of the other planets in our solar system are huge too! For example Jupiter, the largest planet in our solar system, is

1,300 times bigger than the Earth. Saturn is also very large and is 764 times larger than the Earth. At night, with your parents help, you can look at the Moon and planets. You'll have a lot of fun! Trust me!

The solar system and universe were formed a LONG time ago, way before you or me were born. Long ago there was a huge molten rock that exploded. Scientists call this the "Big Bang". Rocks and gas flew everywhere. The rocks and gas formed the solar systems, galaxies, and universe. This took a very long time. Scientists say our universe is between 12 and 14 billion years old. That's really old!!! Our Sun is not quite so old and is about 4 $\frac{1}{2}$ billion years old. Our universe is still expanding to this day from the "Big Bang". Amazing, huh?

Gas formed some of the planets while the rocks formed the rest of the planets. Jupiter, Saturn, Uranus, and Neptune are all gas planets. If you tried to stand on any of these planets you would sink into the gas. There is no solid surface on these planets. Also, another amazing fact is, if you could find an ocean big enough to hold Saturn, it would float! Its density is less than water so it would float.

The gas from the "Big Bang" also formed our Earth's atmosphere.

Isn't learning about space fun? I think it is! Let's continue. Over periods of time, and the planets were done forming, the rest of the rock and gas, in our solar system, formed the Asteroid Belt. This is a bunch of rocks circling the Sun between Jupiter and Saturn. In some very rare cases, these rocks have flown out of the belt and have hit a planet or a moon. Over 1 million years ago, an asteroid hit the Earth.

Scientists think that the impact from the asteroid is what killed off the dinosaurs. But don't worry; scientists don't think any asteroid will hit the Earth for quite some time!

Some of the other rocks and gases formed

comets. Comets are made up of rock and ice and also orbit our Sun. When they get closer to the Sun the ice starts to heat up and it begins to melt. The melted ice causes a vapor that leaves material behind as it travels through space. The Sun reflects on this material and charged particles causing a long tail. So now, if you ever get a chance to see a Comet you will know why it has a tail.

Comets fly by the Earth every once in a while. They are pretty awesome to see! Ask your mom and dad to check online about when the next comet will be in your area.

Our neighbor the Moon (not a planet though) was formed when something crashed into the Earth a long time ago, when it was being formed, causing it to spill out tons of rocks. Not all scientists of space, called astronomers, (as-tron-o-mers) agree on this theory (the-o-ry). Some say a big rock got somehow connected to our Earth's gravity, or pull.

The Moon is a good place to start as an amateur astronomer. It is big and easy to find looking through binoculars or a telescope. A fun fact about the Moon is that it rotates just like the Earth does but we only see one side of it. Why is that? Well, it is because it is spinning at the same rate that it is orbiting the Earth.

Whew! Wasn't that fun! Now you can go tell your friends and family how smart you are! There are many more books that can help you get started. In the back of this book, there is also a resource section where you can look for assistance and more information. There will also be a glossary in case you get stuck on some words.

3
Where are we in the Solar System?

You must be wondering where our planet is exactly in the Solar System. Hmmm... That's a good question!

Well, the Earth is here:

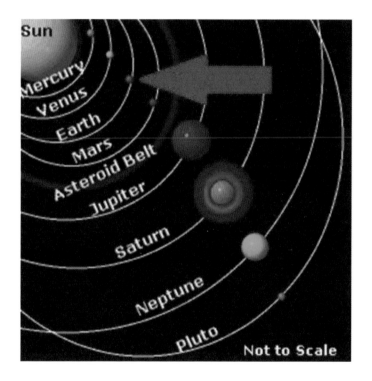

The Earth is in the most perfect position it could possibly be in. It is the third planet from our Sun. If the Earth was a little further away, it would be very cold and we would all freeze!

If we were any closer to the Sun, we would

get too hot and fry up! Life on Earth would not exist if we were any closer or further away from the Sun. You can see how the Sun affects life on Earth. The Sun is very important to us. It provides nutrients to our bodies by its rays, light for us to see in the daytime, and it helps plants grow too. The Sun is a very important neighbor to Earth.

The Earth is 93 million miles away from the Sun. That's far! It takes the light from the Sun approximately 8 minutes to reach us.

If the Sun was blown out, like a candle, we wouldn't know it for 8 minutes. But don't worry, the Sun won't go out for another 5 billion years.

It's incredible to learn what our Solar System is made of! It's a lot of fun learning all about the universe, and how we came to be. In our next chapter, you'll see what astronomers use to look at our Solar System!

4

What do Astronomers use to look at the Solar System?

Hmmmmmm... That's a super, absolutely, positively, the best question I've ever heard! Astronomers use all sorts of tools and gizmos and gadgets to help find things that our eyes can't see.

Let's Begin!!

- **Binoculars** - They are very helpful and easy to use to find objects in the night sky. They are not as powerful as telescopes but you can still see some amazing things with them. Binoculars, like telescopes, can be used for other things too. People use them to do birdwatching, and many other hobbies.

- **Telescopes** - Astronomers use many different types of telescopes to magnify objects in space that our eyes can't see. Like binoculars, telescopes can be used to view many other things also. If you ever seen an old pirate movie they usually use telescopes to look for land or to look for other pirates.

- **Refractor Telescopes** – A refractor telescope uses multiple lenses to bring things into focus. There are usually lenses on each end of the telescope. Usually the viewing lens is adjustable to bring objects into focus.

- **Reflector Telescopes** – Reflector telescopes use, not only lens, but they also use curved mirrors. The mirror captures the light from the object and shines it into the lens for viewing. The larger the mirror the more light that is captured. This allows you to see more things further away in space. There is a telescope in space that scientists control from the Earth to view the universe. It is very large and it also uses mirrors to send pictures back to Earth. It is an amazing telescope and has helped scientist to understand our solar system and universe. With your parent's assistance, you may want to check it out on the Internet. There are some really cool pictures there!

- **Radio Telescopes -** Radio Telescopes use radio waves to look into space. They are very complex telescopes and different then that optical telescopes we just talked about. They use antennas to collect radio waves from space which is used to form pictures. They usually put the antennas far away from populated areas and on mountains so they get less interference from our atmosphere.

- **Maps** -Yes, maps! Celestial maps are used to help scientists keep track of where stars and planets are. They make their lives easier! I use maps a lot to keep track of the stars and galaxies and planets. You can find maps of stars and planets online or in books. Ask your parents to help you find a sky map.

- **Solar Eye Protectors**-very important to protect your eyes from the Sun. They are NOT sunglasses, and can be bought online. It is a lot of fun looking at the Sun! It's awesome! But, again, be very careful and get your parents to assist you. Never, never, never look at the Sun without the proper eye protection!!!

All of these items you can get online. (On Amazon, Ebay, etc.)

I'm hope you've learned a little bit from this book. I hope that you'll be able to look at the universe at night, and not have to wonder what is going on up there.

My fellow astronomer, I wish you the very best of luck and have fun looking at our wonderful, amazing universe!

5

Glossary

- **Solar System** - the Sun and all the planets, satellites, asteroids, meteors, and comets that are all circling our Sun using gravity.

- **Astronomer** – a type of scientist that studies space.

- **Planets** - an astronomical object that orbits a star (Sun) and does not shine with its own light.

- **Gravity** - the attraction, or force, that causes physical things to move towards each other.

- **Our Sun** - the star at the center of our solar system around which Earth and the other planets in our solar system orbit.

- **Star** – A Sun. Like our Sun in our solar system. There are billions of stars in our universe just like ours only further away.

*All of these definitions have been looked up in the <u>Webster's New World Compact Office Dictionary</u>.

6
Resources

1. NASA (National Aeronautics and Space Association)

Website: www.nasa.gov

2. There are lots of other programs online for resourceful uses. I recommend NASA because it has a program for children who would like to learn about space.

7
Bibliography

Aguilar, D. A. (2011). *Planets: The Latest View of the Solar System.* Washington, D.C.: National Geographic Society.

Askew, A. (2012). *The Complete Guide to Space.* New York: Sandy Creek.

Driscoll, M. (2004). *A Child's Introduction to the Night Sky.* New York: Black Dog and Leventhal Publishers, Inc.

Lemonick, J. K. (2012). *Time: New Space Discoveries.* New York: Time Books.

The Usborne Internet-Linked Library of Sciece: Earth and Space. (2002). New York: Scholastic Inc.

Universe: The Definitive Guide. (2012). New York: Dorling Kindersley Limited.

Note: I've read many books, and learned a lot of my knowledge from my father. I've worked hard to make sure I've not plagiarized from any book.

8
Photos and Credits

The fonts used in this book are Comic Sans MS and Calibri, at size 12 to 48.

———————————

I would like to thank my mother and father for supporting me. I thank my mom for helping me edit, and my father for co-authoring this book.

Printed in Great Britain
by Amazon